HARNESS THE WIND
The Story of Windmills

HARNESS THE WIND
The Story of Windmills

JOSEPH E. BROWN and
ANNE ENSIGN BROWN

Illustrated with photographs and diagrams

Dodd, Mead & Company, New York

FOR OUR PARENTS

A tower mill in Northern Ireland

Illustration Credits

The Association Française des Amis des Moulins, 40, 41, 42; Austrian National Tourist Office, 53 (top); Belgian National Tourist Office, 18 (top), 52 (top); Bord Fáilte Photo, 19; Jacques Boulas, 53 (bottom); British Crown Copyright. Science Museum, London, 32; British Tourist Authority, 16, 29, 37, 44; Anne Ensign Brown, 15; Joseph E. Brown, 13, 18 (bottom), 20, 25, 30, 31, 34, 78, 83, 103; Danish National Tourist Office, 10, 51, 104; Dempster Company, 85; Holland Herald Magazine, 59, 62; From *Horizontal Windmills* by R. Wailes, Excerpt Transactions of the Newcomen Society, Vol XL, 1967–68, 21, 46, 47, 49; David Keep, 35; National Aeronautics and Space Administration, 92, 99 (bottom); John Nersesian, 55; Netherlands National Tourist Office, 26, 60, 67, 69, 70, 73; Northern Ireland Tourist Board, 2; Jack N. Oldham, 27; Pennwalt Corporation, 99 (top); Soviet Life Magazine, 52 (bottom); Spanish National Tourist Office, 23, 56; Sun Oil Company, 88.

1 2 3 4 5 6 7 8 9 10

Library of Congress Cataloging in Publication Data
Brown, Joseph E   1929–
    Harness the wind.

    Includes index.
    SUMMARY: Discusses the history, design, and use of windmills throughout the world.
    1. Windmills—History—Juvenile literature.
[1. Windmills—History]   I. Brown, Anne Ensign, joint author. II. Title.
TJ823.B73   621.4'5   77–6489
ISBN 0–396–07484–7

Acknowledgments

Many individuals and organizations aided in the preparation of this book, but special thanks are owed to three: the Windmill Section of the Society for the Preservation of Ancient Buildings, London, England; the Dutch Windmill Society, Amsterdam, The Netherlands; and Mrs. Shelley Hexom.

Contents

HARNESS THE WIND
The Story of Windmills

A smock mill in Denmark

Chapter 1

Windmills and How They Work

Long before the eighteenth-century Industrial Revolution produced power machines to ease human labor, there were machines of another sort that people had learned to master.

They were windmills.

Along with water-driven mills, windmills were the first machines. In fact, they were the *only* machines for centuries, and their origin dates back at least 1,200 years.

Windmills harnessed and put to work a free, natural, nonpolluting, never-ending source of energy—the wind.

To most of us, the wind is a puzzle, a mysterious element of nature which has fascinated mankind since ancient times. Air is in constant motion, and air in motion is wind. That simple definition is as ancient as the early philosophers who first pondered the motivation of this force that covers the face of the earth.

It is difficult for us to understand the wind, because we cannot see it. But if it were visible to us as it passes over the earth's surface, it might resemble a stream flowing along in unpredictable waves and eddies, rushing to its destination in one place, lazily resting in a lull in another, only to be suddenly caught up in a whirlpool the next moment, each twig, rock, or rise of earth interrupting its flow.

Simply put, a windmill is a device that captures this moving, invisible force and transmits the wind's power to machinery designed to perform a specific work task.

Technically, the word "windmill" means a wind-powered machine that grinds or *mills* grain such as wheat or corn. This was the most common function of windmills in Europe before the Industrial Revolution, and a few there still perform that job today.

Throughout history, however, wind machines have aided man in many other ways, though the term *windmill* is used to describe all or most of them. Some produce electricity; more correctly, these should be called wind *generators*. Others pump water and drain flooded areas to make them useful. Still others ventilate mines with fresh air, run foundries, or saw lumber. A few have even been used to produce gunpowder or to extract oil from seeds.

Designed for a specific task, a windmill can perform nearly any job requiring movement that once had to be done by hand: pumping, crushing, grinding, stamping, lifting.

Although windmills were soon outmoded by the arrival of other machines, they had many advantages over their successors. Since the eighteenth century when wind and water were used less and less as energy sources, man has been using up the natural resources of the earth at an ever-

A windmill captures the wind and transmits its power to machinery to perform a specific task. This is a working tower mill in England.

increasing rate. Deposits of coal, oil, and natural gas, the energy reservoir for most modern machines, are dwindling faster than nature can replace them, and the time is coming when new sources must be found.

Coal, oil, and natural gas are called *fossil fuels*. They are *finite*; that is, they have measurable limits. They are produced by growth and decay of vegetation in the earth, a process that takes millions of years.

Compared to the modern machines that fossil fuels run, windmills are weak and inefficient. The "fuel" itself—the wind—is unreliable, unsteady, and somewhat unpredictable. In some places on earth, it flows hardly at all. Before the advent of fossil-fueled machines, these regions were denied machines of any kind; the only reliable energy forms were human muscles and animals harnessed and put to use.

Yet, fickle as it may sometimes be, the wind is a never-ending source of energy to make things go. Unlike fossil fuels, the wind is *infinite*. As long as it blows, there is energy available to make windmills work. For this reason, there are still many in use around the world today, faithfully performing their work just as they have done for centuries.

In America, the most commonly seen type of windmill is a vertical steel structure atop which a rotating, many-bladed propeller catches the wind and transmits its energy to work-performing machinery. The early American colonists built and used a few of the European-type windmills, and the millers operating them were important men in pre-independence America, just as millers were in Europe. But the vast majority of American windmills, adapted to winds in this country, were of the type still seen today on farms

A typical American windmill

and on open prairies where water has to be pumped and electricity generated.

American windmills were simple in construction, compared to European windmills. Thus, they have become fairly standardized. They can be easily taken apart, moved to a different site, and reassembled. Because they are manufactured from the same plans, their parts are interchangeable.

Not so with most windmills in Europe, where no two are alike. Each was designed and built not only with its particular site and wind conditions in mind, but with machinery fashioned for a specific work purpose. Most European

15

The Union Mill dominates the skyline of Cranbrook in Kent, England. One of the loftiest windmills in England, it was restored by the Society for the Protection of Ancient Buildings.

windmills were erected long before the Industrial Revolution. They were so important in the life of rural European villages—grinding grain for flour or pumping water—that they were often installed before other equipment. Many a European village, for instance, possessed a mill long before its parish could boast a clock.

Because of their height, windmills could be seen for long distances and were the major village landmark. Townsfolk often personalized them with scrollwork and individual names. The miller himself—the villager who owned the mill or at least was in charge of its operation—often commanded more authority, power, and respect than the village's elected or appointed chief official.

The miller's task was often a dangerous one, especially when the wind blew hard. It took great skill and strength to prevent a windmill's sails from "running away," damaging or destroying the mill's effectiveness. Too little wind, on the other hand, meant that the sails would not move and the windmill was, for the moment, useless. And with wooden mills, there was the ever-present danger of fire.

The world's earliest windmills, believed dating back to seventh-century Persia, looked nothing like the tall upright structures we see today. They were called "horizontal mills," a term owing to the fact that their arms or *sails* revolved horizontally or parallel to the ground. A few horizontal mills were built in Europe, where the windmill was introduced in the twelfth century. Most there, however, were vertical mills—tall structures on which the sails turned at a right angle to the ground. All American windmills are of the vertical type.

In overall structure, there are three main types of European windmills: *post, tower,* and *smock.*

Above: A post mill in Flanders, Belgium. Below: The underside of a post mill. Mill revolves around the post shown at bottom.

Post mills, the earliest, were always made of wood. "Post" refers to the massive upright post around which the entire body of the mill, that which in turn supported the sails and their machinery, revolved.

The tower mill, often referred to as the Dutch mill, was a development of the fifteenth century. It was built of brick with a timber roof. Its main difference from the post mill was that instead of the entire windmill revolving around a central post, only the top part, or *cap*, revolved.

A small tower mill in Ireland

A smock mill in Rye, England

The smock mill followed. It was a variation of the tower mill. The difference was in construction material. The smock mill usually had a stone base with a wood-frame upper section, covered in weather boarding either tarred or painted white. The name "smock" originated because these mills looked very much like the linen smocks worn at the time.

Regardless of physical differences, all windmills, except

for the horizontal ones, have three major features in common.

First, they must have some means of catching the wind. This is done by *sails*, or arms, that rotate on an axle. Second, there must be a way of turning the sails so that they always face the wind; otherwise, when the wind shifts, the sails would come to a halt. And last, they must have a system of gears and other interlocking equipment that transmits the wind energy to the millstones, water pump, or whatever machinery is used to perform their particular task.

Horizontal mills share in common only the first and last features. Since their sails rotate parallel to the ground, some of the sails are always facing the wind, and there is no need to adjust them, as in the case of upright mills.

The horizontal mill has a simple design. In Europe, they usually consist of six or more sails of wooden boards that

A horizontal mill—sails revolve parallel to the ground

are set upright upon horizontal arms which rest on a tower, and which are attached to a vertical shaft passing through the center of the tower. The sails, in a fixed position, are set obliquely to the direction in which the wind will strike them. As a result, it makes no difference from which direction the wind comes, as it will always blow upon one part of the sails.

The horizontal mill has many disadvantages, and it was never as popular as the vertical mill. The major fault is that only one or two of the sails could catch the wind at any given moment. Since the part of the sails catching the wind must move the "dead weight" of the other sails, horizontal mills are less efficient than vertical mills where the wind's force is distributed evenly on all sails.

To understand how a windmill sail works, think of a bird's wings and sailboat sails. Not all birds fly the same way, or as fast or as efficiently. Birds with long, pointed, relatively flat wings—swallows and gulls, for instance—are the fastest fliers. Sparrows, finches, and other birds with short, blunt, and bulging wings are poor fliers, and chickens, with extremely stubby wings, can hardly fly at all. Early sailing ships had short masts and small sails. Later, yacht designers learned that boats would travel much faster if the masts were raised and the sails made larger.

The same trend of development occurred with windmills. The earliest ones had very short sails. Later, as the sails were made longer, they became far more efficient. Of course, as sails became longer, a means had to be found to place the axle or *windshaft* on which they rotated higher off the ground; thus, the design of both sails and the windmill structure or *buck*, advanced hand in hand.

The earliest sails were wooden frames covered with cloth and were known as *common* sails. At first, the cloth could

This windmill in Spain has cloth sails. It is unusual because of its six arms.

only be placed over the frames or removed entirely, but later a means was found of *furling* or *reefing* the cloth to control the sail area according to the strength of the wind.

Common sails were both light and powerful, but their disadvantage was that they had to be stopped for the miller to furl the cloth. They were stopped by a brake, but if the wind became suddenly too strong before the furling was done, the power of the wind often became greater than that of the brake. Many a mill was lost when the miller waited too long to furl. The danger was that the mill might run out of grain and the millstones run dry; this could cause a shower of sparks that would set the mill on fire.

A second danger of runaway sails was vibration. Especially in early stages of their development, when they were still made of wood, windmills could be shaken to pieces in this way. Facing both risks, the miller often had to ride out a storm just as a sailing ship's captain weathered a storm at sea.

The miller found that one way to slow sails in a high wind was to jam so much grain into the millstones that the stones would slow down and become a sort of brake. A second way was to force the sails edge on to the wind. If the wind suddenly shifted and struck the mill from behind, however, there was a danger that the sails and the windshaft would be blown out of the mill.

In 1772, a Scottish millwright, Andrew Meikle, invented a new type of sail which partially solved the problems. It was known as the *spring* sail. It was composed of a series of shutters, connected by a bar and arranged like a venetian blind, which was regulated by means of a tension spring. The shutter blades opened or closed, depending on the strength of the wind. To adjust the tension of the springs,

Above: Closeup of spring sails on a windmill in England. Left: Springs which control the opening or closing of shutters, depending on wind strength.

the miller had to stop the mill, just as he had done previously.

Each sail was adjusted independently and it was hard work, though not nearly as difficult as struggling with cloth, especially in freezing weather when the cloth became wet and froze.

Most windmills in Europe had four sails like this one in Holland.

The most important improvement in sails occurred in 1807 with the invention of *patent self-reefing* sails by Sir William Cubitt. These were similar to spring sails, except that the opening and closing of the shutters could be controlled automatically by a weight suspended outside the mill. This weight was connected to the shutter bar by a series of rods linked to a lever called a *spider*. No longer did the miller have to halt his grinding of grain when the wind blew too hard; the adjustments could be made while the sails were turning.

The function of the shutters on both spring and patent sails is to allow part of the wind to pass through them in

A post mill in Bulgaria. Note the unusual number of sails.

heavy wind. When the wind is light, springs close the shutters to gain the best use of the wind. The tension of the springs can also be adjusted.

There was constant experimentation with materials used in sail construction. Wood, of course, is lighter than metal and was a favorite. However, it is not as strong as metal, so on some windmills, a combination of the two was used: metal for the outside frame and wood for the shutters themselves.

Although most European mills had four sails, some operated with only two, and it is believed this practice originated when a storm broke one or two sails on a four-sail

27

mill. On the other hand, there were mills with as many as eight sails. Regardless of the number of sails, they always rotated in a counter-clockwise direction, as the windmill is viewed from the front.

How long sails would last depended upon how often the mill was used, the wind conditions for its particular area, the quality of materials used in its construction, and other factors. Although many sails lasted forty or fifty years, replacing them at frequent intervals was far more common.

A few examples of the early common-sailed windmills exist in Europe today. Most of them are mills that have been restored for historical purposes, since in most cases spring and patent sails replaced the cloth types.

Keeping the sails pointed toward the wind was at first an operation done by hand. Most older post mills, for instance, were fitted with a *tail pole* extending to the ground on the side of the buck opposite the sails. When the wind shifted, the miller turned the mill by pushing the tail pole.

An invention of the mid-1700s at last gave the miller a respite in this part of his daily labors. Called the *fantail*, it was a device which kept the main sails pointed automatically into the wind. The fantail is itself a small wind-driven sail, but it faces the opposite direction from the main sails, and is located on the back side of the buck. The fantail is really a windmill in miniature; if the wind changes direction and blows against the side of the mill, it catches the fantail and makes the mill start to turn. As the fantail is blown to an idle position, the main sails are pushed into their correct position again. In the case of post mills, the mill is turned on a central post. On smock and tower mills, only the top part, or *cap*, of the mill revolves. The connection between the fantail and the movable part of the mill is

28

A post mill at Saxtead, England. Note fantail in foreground (back of mill) which rotates the mill on wheels.

a system of gears which moves very slowly in relationship to the speed of the fantail.

A safety device was fitted to most mills to disengage the fantail when it became damaged. Turning was then done by a hand crank until the fantail was repaired.

Keeping the fantail in good running order was another dangerous job of the hard-working miller. In most cases, it was placed near the top of the mill and very far out for the greatest efficiency. To lubricate its gears or to replace a part meant crawling out on a narrow bar, although later fantails were connected with safer walkways.

The principle of the fantail applies to both European and American windmills, except that in the latter case the fantail is simply a vane, similar to a weather vane. The flat-sided vane is mounted vertically. As the wind shifts, it strikes the side of the vane obliquely, pushing the main blades back into position.

Fantail, cap, and part of the sails on a tower mill in Packenham, England. Fantail keeps main sails pointed toward wind.

Closeup of gears and wooden gear teeth in an English windmill

The *windshaft* is a main unit which connects the sails to the mill's internal machinery. On early mills, it was octagonal or round, and tapered from the breast or front of the mill to the tail, or rear part of the mill. Because it had to transmit the force of the wind to the gears and other machinery that operated the millstones, it was made of sturdy wood or iron.

The inner end of the windshaft is fitted with the *brake wheel*. The largest wheel in the mill, it sometimes measures ten feet across. *Gear teeth* are placed around the rim; they drive the *stone nut* which in turn drives the millstones. Around the perimeter of the wheel is the *brake* which consists of a ring of sections of wood which almost encircles the brake wheel.

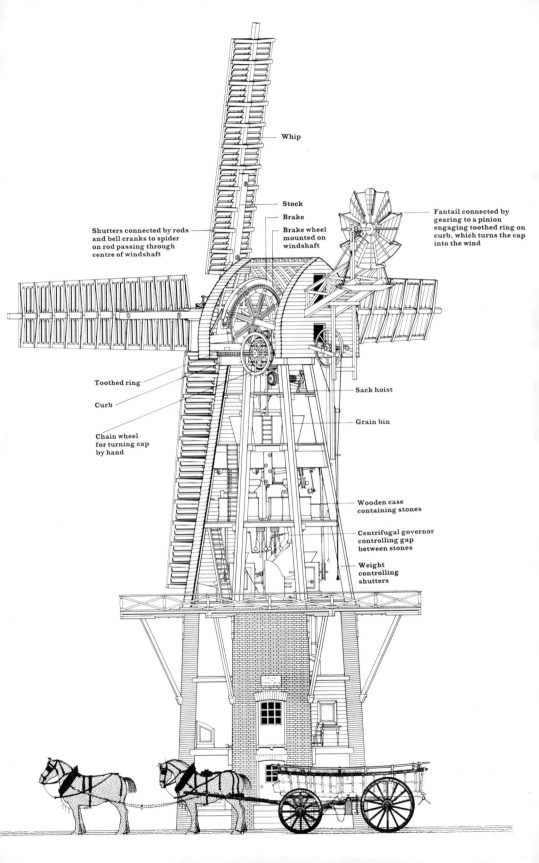

Whip

Stock

Brake

Brake wheel
mounted on
windshaft

Shutters connected by rods
and bell cranks to spider
on rod passing through
centre of windshaft

Fantail connected by
gearing to a pinion
engaging toothed ring on
curb, which turns the cap
into the wind

Toothed ring

Curb

Chain wheel
for turning cap
by hand

Sack hoist

Grain bin

Wooden case
containing stones

Centrifugal governor
controlling gap
between stones

Weight
controlling
shutters

The windmill brake operates something like the brake on a bicycle wheel. Operated by a lever or rope, it simply clamps itself down over a large section of the moving brake wheel and slows it down or stops it completely. Thus the miller has two ways at his disposal of regulating the speed of the sails: by changing the sail shutters to increase or decrease the amount of wind passing through and by braking the wheel.

The energy which has come from the sails through the windshaft to the brake wheel is now transmitted to other gears which in turn operate the millstones. Careful planning is necessary in the design of the various gears. Since friction of metal to metal or wood to wood can cause fire, the material used is usually alternated. In other words, metal gears mesh with wood gears instead of to their own material.

The millstones comprise the working hub of any grain mill. There are two millstones; the lower or *bedstone* is stationary, the upper or *runner stone* rotates. Because of the small size of the grain and because the millstones were to crack and mill the grain but not crush it, the stones had to be very carefully made and the machinery that ran them able to adjust the distance between them to a very fine tolerance.

The grinding surface of the stones has a number of grooves cut in it which convey the flour to the periphery. On each *land* between the furrows are cut several fine grooves which carry out the actual grinding. There are various methods for adjusting the distance between the stones. But since even a change of one thousandth of an inch in the space between could change the quality of the flour itself, the adjusting machinery had to be built so that

Mill grindstone and various hoppers from an English windmill

many turns of a screw or lever would adjust the stones only minutely.

Often, the reputation of a miller depended partly on how closely he could adjust his stones. The experienced miller learned to tell by feeling the flour whether his adjustments were right; if they were not, he continued feeling the flour while he turned the adjusting screw to find the best setting.

In early mills, the miller was busy adjusting his stones practically all day, for as their speed increased, the whirling force of the stones constantly widened the gap. Even a momentary burst of wind could change the gap just enough that flour quality was altered.

To avoid this trouble, later mills were fitted with *governors*. They were usually belt-driven weights which centrifugal force pushed out from their base as speed increased, and adjusted the stones automatically.

Over a period of time, the grinding action wore down the stones' grinding surfaces, and the stones had to be removed for regrooving, or *dressing*.

To prepare the grain for grinding, the miller used a *sack hoist*. The simplest form was a long rope or chain that pulled the grain sacks from the floor of the windmill, over a pulley system, to some point higher than the millstones. On later mills, machinery was devised to utilize the mill's wind energy to hoist the grain; on early mills, however, it often took a strong back and arms, as this was accomplished by manpower alone. From the location of the sack hoist, the grain is fed by gravity to a *hopper* on the millstone level of the windmill. After grinding, the grain is carried to a grain

Grain hopper of Draper's Mill, Margate, Kent, England

floor below, where it is sacked and taken from the mill, ready for transport.

Though in principle they operated generally the same as wheat and corn mills, barley mills and rice-hulling mills differed in some detail. Instead of milling stones they used hulling stones; somewhat larger than the common mill-stones, their casings were constructed a little differently. Barley and rice does not have to be ground; it is only *hulled*, that is, the thin outer covering needs only to be removed before its use.

Hulling stones were usually gritstones. They had only a few deep, wide furrows, and the grain was flung out along them without being ground.

Because they required heavier machinery than grinding mills, hulling mills also needed more wind. The places in Europe where they could be erected were thus more limited.

There were many other types of windmills and each had machinery specially suited to its operation. There were oil mills for pressing oil from seeds, mustard mills, as well as mills in which many kinds of coarse and hard materials were ground fine. The grindstone or crushing device was the piece of equipment common to all of them, even though they varied considerably in design, placement, and operation. Although windmills were carefully planned in advance, there was really no way to tell how efficiently they would work until they were erected and put into operation. Sometimes, unforeseen problems developed.

One example involved two mills in County Suffolk, England, one a smock mill and the other a post mill. Built fairly close together, it was not discovered until they were in operation that one tended to "rob" the other of wind

King's Mill, a smock mill at Shipley, Sussex, England, restored and opened to the public.

when its sails were revolving. In another case, a miller in Kent, England, once went to court trying to prove that a shed built by a railroad nearby had created an adverse wind turbulence on his mill. The miller claimed that the situation was so bad that the turbulence broke a sail. No one could see the wind, of course, so the miller found an ingenious way of proving his point. He built a bonfire near the shed, on the windward side of the mill, started his windmill, and pointed out how the smoke curled and "bent" as it floated toward the sails. He lost his case nevertheless and solved his problem only by raising his mill higher on another base.

Many wooden windmills were moved when it was felt they would perform better in a new location. With the heavy moving equipment today, moving a windmill would hardly be a problem. Centuries ago it was a tremendous task. Some had to be entirely dismantled and moved piece by piece. Others were moved in their entirety. Perhaps the most famous windmill moving involved that of a post mill in Sussex, England, that required several teams of oxen which dragged the mill on tracks that were placed along the route.

Today, farmers are able to buy grain for their livestock and bakeries flour for making their bread from big milling companies whose equipment is tremendously more powerful and productive than in windmills of the past. A few drainage mills and water-pumping mills continue to operate in Europe, and in dry areas of the United States there are still thousands of water-pumping windmills at work. Except for these, however, and the few windmills that have been restored, there is little remaining evidence of man's ingenuity in developing his first machines.

38

Chapter 2

Windmills in History

The early miller, like the ship's captain, led an arduous, exciting, and often perilous life, depending upon the fickle wind for his livelihood.

Pulling down the sails, spreading the cloths, then releasing the brake, the whole mill trembled, breathed, and came to life as the sails began to turn. The low murmur of air rushing past the sails, the familiar click of the wooden gears, the whispering stones, and the rustle of grain filling the hopper were music to the miller's ears as he toiled on in the mist of the rising flour dust.

These were the good times, but he knew that the very force that drove his mill could also destroy it. There could be days on end when barely a breath of wind stirred, stopping the stones completely. The piles of unground grain would accumulate higher and higher as the flour mist settled in the silent mill.

As frustrating as the silence was the howl of a brewing storm. As the miller prayed for a light wind, a sudden gale might spring up and the mill would groan and sway dangerously. Working desperately to save his mill, he would

Views of a modern-day working grain mill in France. Grain arriving at the mill.

Below: Putting the cloth in place on the sails

Rear view of the windmill in France. Note the tailpole which brings the sails into the wind.

Below: The bed and staircase to the second floor of the mill

Interior views of the French mill

wish again for the silence to replace the menacing wail of the wind that tossed the mill about like an overrigged ship, "tailwinding" the mill, lifting the cap to the winds, and often even dashing the mill to final ruin.

From olden times to today, the wind has been the constant, common factor that has governed the development of windmills. As man became more civilized, discovered new metals and building materials, and as his engineering skills improved, he continued to increase the efficiency and usefulness of windmills. But every step of the way, it was the wind that guided his accomplishments.

At first, the wind was understood only in terms of its destructiveness. Even today, hurricanes, tornadoes, typhoons, and other storms cause great damage and loss of life around the world. Out of fear and respect for the wind's awesome power, legends developed, suggesting that the winds were imprisoned underground and released at the whim of a malevolent or beneficent deity. To the ancient Greeks, this god was Aeolus, ruler of the winds; many a legend exists telling of his caprices with "his" winds.

A belief that the wind could be controlled by magic was also common among many primitive peoples. Sometimes power over the wind was ascribed to royalty. To the early Christian, it was thought that the wind was God breathing to punish or reward as he saw fit. And then there are the many legends among centuries of mariners dealing with fear and superstition involving the wind.

Man finally overcame his fear, and this power, so long cloaked in awesome veneration and superstition, was looked upon as a challenge. Just as man once learned to put fire to good use for cooking his food and warming his home, he also began to wonder if the wind might be used for his

Wind was first used for navigation by sail.

benefit. By constantly observing the whimsies of nature's forces the idea of harnessing the wind and water evolved. It was found that a wheel could be whirled to move objects, a sail filled with wind to create power.

Wind was recognized as a means of reducing the effort of rowing as early as 4000 B.C. Fragments of Egyptian pottery from this era depict ships with square sails navigated by the wise Egyptians who took advantage of the prevailing north wind up the Nile. By the Egyptian Fifth Dynasty, about 2500 B.C., sailing ships had advanced enough to go to sea, but only as auxiliaries. Slaves were cheap and they remained the prime mover at sea.

As the sail persisted, other important inventions utilizing the wind were being made. As early as 150 B.C. an account is given in *Spiritalia*, the historic manuscripts by Hero of

Alexandria, in which a wheel driven by the wind worked the piston of an air pump to blow the pipes of an organ. The ancestor of the Eastern or horizontal windmill was perhaps the prayer wheel used in central Asia around A.D. 400 in which scoops that caught the wind rotated on a vertical shaft.

What was probably the first real windmill was invented in Persia during the reign of Caliph Omar I, when a Persian subject claiming to be able to build a mill rotated by the wind was asked by the Caliph to do so. That first Persian millwright was granted a charter in the year A.D. 644.

The next account of windmills comes from two Persian geographers. The writing tells of windmills in the windy and sandy province of Seistan, Persia, around A.D. 950. The people of Seistan first harnessed the winds to control the sands which had drifted to such heights that, if means were not taken to control them, entire cities could disappear. To remove these drifts and spread the sands in nearby fields, they enclosed the drift or dune in a structure of timber and thorns higher than the dune. In the lower part they made a door for the wind to enter and blow away the upper levels of sand like a whirlwind, tossing the sand to fall evenly over the surrounding area.

At about this same time we learn that there were also windmills that pumped water. An Arab geographer, Ali al-Mas udi, writes of Seistan in A.D. 947 that ". . . wind turns mills which pump water from wells to irrigate the gardens. There is no place on earth where people make more use of the wind."

Little more than three centuries later, in A.D. 1283, al-Qazwini, a Persian scholar, described how the people of Seistan used the wind to grind their corn, as well as to

Line of mills at Seistan

control the drifting sands and pump their water. From these early writings many scholars believe Seistan to be the birthplace of the Eastern horizontal windmill in early Muslim or even pre-Muslim times.

Detail of Seistan mill

Most of the windmills we see today are built vertically; that is, their arms or sails are mounted on a platform and turn in the wind at an angle nearly perpendicular to the ground. The machinery that the sails move (such as grindstones) is always located below the level of the sails and the shaft on which the sails turn.

Many early mills, however, were of the horizontal type and they differed from vertical mills in two important respects. First their arms rotated parallel to the ground. One such mill was the Persian windmill, described by the Syrian cosmographer al-Dimashqi (A.D. 1300). Generally erected in a high place such as the top of a minaret or tower in a castle, the mill was built as a two-story structure. The upper story contained the mill that turned and ground; the lower part housed a wheel rotated by the captured wind. (The wheel was not called a sail as in later vertical windmills.) This wheel turned one of two millstones in the

Twin corn mill at Seistan, Iran

upper section. The fact that the millstones were *above* the wheel made it different from the vertical mill in the second major respect.

Four slits were made in the walls of the first story. The outer part of the slit was wide, the inside made narrower, forming ducts for the wind, from any direction, to penetrate the interior with great force. This wind force then hit a reel with six to twelve fabric-covered arms. As the wind filled the fabric sails the arms were moved forward, turning the reel, and the motion of the reel moved the millstone which ground the corn.

Similar horizontal windmills were used in Afghanistan almost as early in those in Persia. These mills were driven by the north wind. Attached to the windmills was a series of shutters which opened and closed to admit or keep out the wind. These windmills evidently were a local Persian adaptation of the horizontal Greek watermill (which originated in northwest Persia) to a country where there was little water but where steady winds prevailed.

In use for centuries only in Persia and Afghanistan, windmills began to spread throughout the Islamic world and on to the Far East to grind corn, pump water, and crush sugarcane.

Because of widespread devastation by invading Mongols and Turks, there are few of these old windmills left in the Near East, except in Persia. There, a few are still performing their faithful tasks of irrigating farmlands and pumping salt brine.

Few horizontal sugarcane windmills remain in Egypt today, but the Egyptian mill lives on far away in the West Indies of the Caribbean. Centuries after the windmill's usefulness in Egypt, it appeared in the Caribbean when

48

Chinese mill

West Indians brought Egyptians in to help establish the first sugar plantations there.

In the twelfth century, when Genghis Khan's hordes swept through Persia, captured millwrights were taken back to China where horizontal mills rigged with sails of matting became common. This type of mill, derived from the Persian windmill, is not in a millhouse but stands vulnerably in the open fields. Its sails, adapted from Chinese fishing boats, are attached to a circular bamboo frame. There are comparatively few of these mills, as it is said that superstition prohibits their use.

Egypt waned, Greece was toppled by Rome, Persia fell before the Mongols and Turks, and the unchanged Eastern horizontal windmill struggled for survival.

Meanwhile, windmills began to dot the meadows of Western Europe. As no written records exist, there has been great speculation on the origin of the Western or vertical windmill. Some historians believe that the concept came from Persia by way of the trade routes through Russia and Scandinavia. Others think it may have come from Islam by way of Morocco and Spain. On the other hand, the vertical construction of the early Western post mills suggests that they may have been derived from the Vitruvian watermill, a vertical Roman watermill which it closely resembles. Or perhaps the vertical windmill was independently invented in Western Europe. In any case, from the twelfth century on, the construction and use of windmills spread throughout Europe.

The first documented information on the existence of the windmill in Europe comes from Mabillon, an early scholar. In A.D. 1105, he writes, a convent in France was allowed to erect watermills and windmills. According to Mabillon, windmills were becoming common in Italy by the twelfth century. Questions arose as to whether the tithes to them belonged to the clergy. Pope Celestine III decided the controversy in favor of the church. Mabillon also writes of an abbey in Northamptonshire, England, in the year A.D. 1143. The abbey was situated in a wood which over a period of 180 years was destroyed. The cause of destruction was said to be ". . . that in the whole neighborhood there was no house, wind or water mill built for which timber was not taken from this wood."

In England the first definitely known windmill was built in A.D. 1191. It was located in Bury St. Edmunds in defiance of vested authority and consequently was destroyed by the abbot. Like many desired but forbidden things, windmills rapidly multiplied. By the fourteenth century they were plentiful enough to permit the British monarchs and other high officials to watch their victories and defeats in battle from the tops of windmills.

50

A tower mill in Denmark

In A.D. 1222, a windmill was erected on the town wall in the city of Cologne, Germany. In 1237, the first definite documentation of windmills in Italy appears, describing a windmill at Siena. The oldest Dutch record tells of Count Floris V granting the burghers of Haarlem the right of paying six shillings tax for a windmill and three for a horse-mill in the year 1274. This amounted to a "tax reduction," since noncitizens were charged at a higher rate for the privilege of operating a mill. In 1299, a windmill was built at Koningsveld Monastery at Delft. The building of a corn mill at Logehem was found in the accounts of the Count of Gelre for the year 1294.

From the twelfth to the fifteenth centuries the construction and use of windmills spread over Central Europe north through Scandinavia. They reached Finland about

A tower mill in Belgium

Windmill built in the nineteenth century in Latvia

Grapes are grown and harvested near this windmill in Austria.

Windmills in France

A.D. 1400, spread east from England through the Low Countries and the North German plain to Latvia and Russia and southward. There, they were common enough by the close of the sixteenth century for Cervantes to feel sure that no one would wonder what his hero, the knight Don Quixote, was tilting at on his encounter with windmills on a Spanish plain.

In the south the quaint windmills of Crete and the Aegean Islands, with examples as far west as Portugal, must not be overlooked. The Cretan windmills are mysterious in origin; they are tower mills, but all other tower mills evolved from the post to the tower stage, and yet the post mill has never been recorded as existing in the Aegean area.

Manuscripts of the thirteenth and fourteenth centuries do not show jib sails (triangular sails) which are found only in the Mediterranean and Iberian Peninsula. These factors led historians to believe that these windmills were probably early adaptations of the Western tower mill to the unique conditions of the Mediterranean. Here the windmill's similarity to the sailing ship again appears. Unlike other instances in which the windmill patterned itself after the sailing ship, this was the reverse. These early Cretan windmills were structurally and aerodynamically an ancestor of the modern racing yacht. This mill was a triangulated structure flying six to twelve triangular canvas sails. These sails, fully capable of being put up and down by rolling the sail around its support or *roller reefing*, are situated at the center of the mill or "sheeted amidship" and the mill sails on and on with the wind on the beam. The windmills of the Aegean Islands were small compared to West European standards, ranging from only four to twelve meters in diameter, while the Dutch polder mill

Note the triangular sails on this windmill in Greece

measured approximately twenty-nine meters in diameter. Four, eight, or twelve little windmills in a row may well be better than one large one, as the use of wind power in Crete survived long after its decline elsewhere. This may be attributed not only to the local circumstances but to the superior efficiency of these picturesque little inventions.

As the windmill became the typical prime mover all over Europe for grinding grain and corn, pumping water, papermaking, pressing oil from seeds, grinding pigments and chalk, sawing wood, and even flinging beehives into besieged towns, a new type of windmill was developing in Holland. It was the Dutch drainage mill, the only means

Three windmills in Ciudad Province, Spain, show how windmills often work in tandem.

the early Hollanders had of keeping dry the land reclaimed from the sea. At one time, more than two thousand windmills drained some 2 million acres.

It is thought that the Cistercian monks of the Cistercian order in France introduced this type of windmill to drain the lakes and fens of their region. The production of peat for fuel for the larger settlements formed shallow lakes which grew through the centuries, finally requiring systematic drainage by the 1300s. The first Dutch marsh mills

56

were built around 1400; by 1600 they were a common fea-
ture of the Dutch drainage system. The steam pump finally
replaced the mill, for though the wind is free it is also
whimsical and often refused service when it was most
needed.

With the harnessing of steam, the windmill, like the
commercial sailing ship, dwindled considerably in use.
Both ship and windmill were important to man during
their time, but steam was more efficient. The age of steam
not only caused the windmill and sailing ship to lose their
importance, but saw a change in the lives of people involved
with both. Certain job skills were no longer required, for
instance, while new ones were created in their place. But
though both sailing ships and windmills are nearly gone, it
is important to remember the vital role they played in our
history.

Chapter 3

The Nation That Windmills Built

In The Netherlands there is an old proverb that goes, "God made other countries, but the Dutch made Holland." Though this may sound a bit presumptuous on the part of the proud Dutch, it seems likely that even if God did start the creation of this Low Country of Europe, the Dutch *finished* it.

In the early Middle Ages all of Western Holland was only an uninhabitable chain of lakes, peat bogs, and salt marshes. That area now includes Holland's major cities of Amsterdam, Rotterdam, and The Hague. Today, Holland is a small but prosperous country of 12 million citizens where every inch of land is still considered precious because there is so little of it.

The land, mostly below sea level, was separated from the sea only by coastal sand dunes. Often in spring the high tides, along with the northerly gales, would cause the sea level to rise far above its normal height, flooding the lacework of land that enclosed the lakes and marshes.

The early Dutch wrested this land from the angry seas. At ebb tide, with only their hands and their tenacious na-

Windmill models are a popular hobby in Holland.

ture, they built dikes around the land and then gravely defended them from the incoming tide, adding earth and stones and endlessly scooping out the trapped water. Later the scooping was done by ox-powered bucket wheels, but even this was tedious; no sooner was the land drained than it was again inundated by the ceaseless tides.

This flat country had no fast-flowing streams for power as did many countries. It did, however, have plenty of wind. So the Dutch turned to the wind to drain their lands.

At first, the wind was an enemy, screaming over the flat wasted bogs, whipping up the sea for attack on the sandy dunes, and causing endless toil on the dikes. Then the Dutch turned these powerful winds away from destruction by erecting windmills behind the newly built dikes, catching the wind in the sails, animating the mills, and trans-

59

Mills at Zaandam, Holland

forming the lakes and marshes into fertile tracts of land, called *polders*.

These mills were driven by four sails attached to a hood. As the sails turned they moved a scoop wheel which lifted the water from the diked-in polders and dumped the water into a canal. As the direction of the wind changed so changed the hood, therefore keeping the mill in operation as long as the wind would blow. And so it was the dike and the windmill that actually enabled the Dutch to create their country, winning the land from the sea.

The windmill in the Dutch landscape seems as if it had sprouted up quite naturally from the land in perfect harmony with its natural surroundings. Planted solidly on the earth, exuding energy, all of its primitive parts impart simplicity, realism, and practicality, mirroring the gravity of the Dutch character, at one with and a part of nature's wind and water.

Seeing a Dutch mill rooted in a pasture crisscrossed by polder channels with the whisper of wind in the sails, one has the feeling that it is almost human, with both the strength and frailties of man. Maybe that's why in the early days when the many mills were alive with the breath of wind, they were considered to be members of the miller's family, sometimes even sharing the family name.

The mill was shaped and clothed just as were other members of the miller's family: it was wrapped in a coat or pelt (weather boarding), and had a ruff (a circular-shaped wooden board round the neck of its axis), and it wore a cap or coif. It even had a bib like the miller's children, bearing the year in which the mill was built.

The windmill had its daily task to perform just as the other members of the miller's family had theirs. These tasks varied, depending on the time or century, the location and the type of mill.

Every mill was christened. Many had women's names such as "The Maiden," "The Princess," "The Housewife," or "The Empress." Some were like the family pet: "The Grey Goose," "The Falcon," or "The Iron Hog." Biblical names resounded too; there was a "Joshua," an "Abraham's Offering," and a "Bethlehem."

But, like people, some windmills were given nicknames which were remembered long after the christened name was forgotten. "Joshua" became "Fatty," and "Abraham's Offering" became "The Blind Ass." If the miller had a philosophical turn of mind the mill might be "The Hope," "The Future," or "Father's Blessing." Drainage mills had their own unique names befitting their task, like "The

Dolphin," "The Drinker," or "The Splasher." It is strange that although windmills in other countries are always "she," in Holland, though given women's names, they are referred to by the miller, using the personal pronoun of the opposite sex. "He turns his face to the master," the miller might say of a mill in "mourning."

The mill being one of the family was shown by its participation in their sorrows and joys. Setting the sails in certain ways could give various information, indicate particular circumstances, and express certain sentiments. Like the burgomaster, the schoolmaster, and the notary public, the miller had an important position in the rural community.

Everything that went on in the village was of interest to the miller, who discussed the news and gossiped with his customers. Through the language of the sails he passed the word along to the villagers. It was quite simple to read the "code" of the sails.

When the miller wished to express joy he moved the sails so that one of them stopped just before it reached the highest vertical position. This was called the "coming position" and announced the birth of a son or daughter, a birthday, or a marriage. Often, especially in the Zaan district, joy was expressed not only by the position of the sails alone but by festooning the entire mill with finery, particularly for weddings. Simple works of folk art, small flags, cutouts of hearts, trumpeting angels, wreaths, cupids with arrows, and other festive symbols in glistening foil and bright-colored paper hung all about the mill. Sometimes an extra sail was laced in and out of the framework, reminding one of a village square where a fair was being held.

When the upper sail was fixed just past the highest position it was in the "going position," or the position of mourning. This indicated a death or some other sad occasion. The body of the mill was turned toward the house of the deceased or unfortunate. If a funeral procession passed by, the mill, with sails set in the mourning position, turned its cap and sails to follow the procession, stopping when it ended, facing the village churchyard. It then remained in mourning for six weeks—unless there was too much work to be done—silently sharing the grief.

Other messages were conveyed by the sails, such as telling the miller's assistant to return quickly to the mill by placing the arms vertically with all sails set except for the one on the lowest arm. There were signals to send for a millwright to come for repairs, or information that the stones were being dressed, so that for some time to come no grain could be accepted for grinding. Position of the sails on a drainage mill gave the signal to start or stop pumping in connection with the level of the water of the storage basin, the water level in the polder, or the fouling of channels and ditches.

At the time of the Reformation, Catholics were being persecuted in the northern provinces and services were banned. Catholic millers sent secret messages by the wings on their mills, telling when and where Mass would be held.

In the dark years of World War II the miller used the "sail code" to convey the latest good news. During the German occupation this code was used to send messages to members of the underground forces and to resistance fighters or pilots of allied aircraft. If the miller was caught, the penalty often was his life.

The position of the sails were also meaningful in day-to-day life. When windmills were in the "rest" position the sails formed a cross. This was the proper time for various maintenance work and a fresh coat of paint.

When the sails were at an angle of 45 degrees to the vertical, it was an indication that the mill would be unused for a long period of time. However, when a drainage mill's sails were set in this manner it was said to be in the "protest position," showing that though the mill was ready for work it could not continue its task due to too much vegetation growth in the drainage channels. A rest of short duration during the milling season or "as soon as the wind comes up" was shown by the "working position," one pair of sails in the vertical and one pair in the horizontal position.

In 1579 the various Dutch provinces entered into a federation forming "The United Provinces" which increased the prosperity of the country greatly. The East India Company was formed and in every direction new trade routes were opened. The Dutch sent their sailing ships across the ocean to distant countries, bringing back goods from these colonies to process and sell in Holland.

Just as the Dutch availed themselves of the wind to sail their ships, they also relied on the wind to process the goods by using the windmill. Although mills for grinding corn had existed in Holland from about 1200 on, it wasn't until the 1500s that other industrial processing mills appeared.

Mills for extracting oil were first erected in 1582. In 1586 the first paper mill appeared, and a mill for sawing wood for building ships and houses was constructed in 1592. Then mills for other tasks began springing up everywhere as the raw materials arrived from the colonies for processing. There were mills for spices, textiles, cocoa, mustard, rice, and barley. There were lime mills, fulling mills (for treatment of cloth), and tan mills for grinding oak bark for tanneries. There were even mills for grating tobacco for snuff and making *jenever* (Dutch gin), Holland's national drink. Then, of course, there were the unique drainage mills dating from the early 1400s.

With shipping and the large-scale industrial processing of products, along with the acquisition of new fertile land, an unparalleled prosperity arose which, owing chiefly to the windmill, led to Holland's "Golden Age."

The various types of Dutch windmills can be classified by both their outward appearance and the task they performed.

The oldest type mill, found mainly in the southern part of Holland, is the *standermolen* or post mill. These curious wooden corn mills of considerable antiquity are still seen in Flanders and are the true archtype of the Dutch windmill. With this type, the whole body turns about a pillar which extends directly into the ground. If the base is completely enclosed it is called a post mill with a round house; if it is open, with the cross trees and quarterbars or brick piers visible, then it is called an open post mill.

From these ancient post mills the *wipmolen* or hollow post mill was developed. The wip mill is chiefly a drainage mill with only a few functioning as corn mills. These graceful mills are still to be found in south Holland and Friesland. In the *wipmolen* the whole upper part revolves around a hollow post to set the mill into the eye of the wind. This is held in its vertical position by a pyramidal base. The upright shaft extends through the hollow post. As the shaft rotates, water is drawn up and then discharged through its upper end. The wip mill, like the post mill, has a tail pole, ladder, and winch. Wip mills are usually red, blue, green, or yellow, trimmed in a white frame with a brightly colored star on the pole end.

When the need for larger and more powerful windmills arose, the *bovenkruier* (upper winder)—the large Dutch drainage or polder mill—was developed. These mills appeared in the watery region of north Holland in the sixteenth century. They evolved from the wip mill, with a smaller top and a larger body, in order to generate more power but still remain manageable. They have an octagonal brick base and a thatched body with a movable cap turned by a tail and capstan wheel.

To raise the water, a scoop wheel was used for shallow

Post mill in Holland. On this mill, the post is covered by a lower structure.

polders and a water screw for deep polders or lakes. A number of these larger mills were built in predetermined places along a canal called a *ring canal* which encircles the lakes to accommodate the drained-off water. The water is then carried to large rivers and then to the sea. Further development of these large mills was directed toward turning the cap from below by a tail pole, and thus arose the familiar large mill of south Holland, the *buitenkruier* (outside winder). The unchanged early north Holland type has been preserved as the *binnenkruier* (inside winder). This mill is heavier, has a base of wood, no cap, and is turned from inside, but it still holds its own in north Holland today despite its handling with less ease than its southern counterpart.

Aside from the large polder mills and wip mills there are smaller varieties of each. One of these small mills found in Friesland is known as the *spinnekop* (spider). The construction and workings of the *spinnekop* are quite similar to the larger wip mill but it has a smaller capacity and is equipped with an open Archimedean screw.

Another very small mill is the *weidemolentje* (meadow mill), sometimes called *aanbrengertje*, a small drainage mill of the wip type used as an additional pumping mill. This diminutive meadow mill of north Holland pumps the water into small polder ditches rather than into the main ring canal. A small flat wind vane at the back of the movable top keeps it always into the eye of the wind. These mills are usually only ten to twelve feet high with a three- or four-bladed wooden pump to raise the water.

Also, there is the primitive *tjasker*, a small simple mill found only in Friesland. It consists of an inclined shaft, the upper part carrying a set of tiny sails with boards and cloth,

Tower mill in Holland. Note circular platform to reach sails.

the bottom of the shaft ending in the Archimedean screw immersed in a deep pond. The trestle rests on a circular wooden beam lying loosely in a ring of concrete round the pond. The trestle turns on this ring to face the wind, the pole in the center of the pond acting as the pivot. There are very few of these left, and all are now in disuse.

A windmill placed in a village among trees and other buildings must be tall to catch the wind. Corn mills are usually located in these cramped quarters. With the need of large working and storage areas, the Dutch millwrights of around 1640 began building large tower mills or *stelling-molen*. To make it possible to operate the mill a stage was built round the mill about halfway up. On the ground

69

floor there was ample room for horses and carts to transport the grain from customer to miller. Before these tower mills came into use mills were placed on the city walls on the outskirts of town. These were called *walmolens*.

In the higher regions of eastern and southern Netherlands a mill similar to the *walmolen* were built on artificial mounds. These *bergmolens*, sometimes called *beltmolens*, can be operated on the circular wall of earth surrounding it just as the stage is used on the *stellingmolen*. Parts of the mound were dug away to make space for the grain to be brought to and from the grinding area.

In the Zaan district, where timber arrived from various parts of the world by ship, the sawmill yards and timber trade were established. Thus these wooden sawmills as a

A wind-powered sawmill in North Holland

rule were situated on the waterfront, sometimes completely surrounded by water. Such a mill was built to turn as a whole to face the wind. On either side of the mill, sheds or wings were built, closed in front and open in the back, to accommodate the long logs to be stored for sawing. This gave a curious appearance to the mill to which it owes its name. It is called a *paltrokmolen*, from its likeness to the flaring coats worn by the *Pfaltzrok* (Mennonites) from Pfaltz who emigrated to Holland for reasons of religious persecution.

The first known sawmill was *Het Juffertje* (The Damsel), built by Cornelis Cornelisz of Uitgeest in 1592. It was later moved by wooden raft in 1596 to Zaandam. It was from this first sawmill that the *paltrok* developed about 1600, followed by sawmills of the smock-mill type.

The sawmill of the smock type is not octagonal as the other types. Low sheds extend beyond the wings as part of a *paltrok* ending with a slipway into the water so the logs can be dragged along by the wind from the water to the saws. Most of the other equipment is similar to that of the *paltrok*.

Finally, there were a few small smock and wip sawmills, sometimes with circular saws, used for the lighter work. The smallest of these, the *lattenzagertje*, a wip mill perched on a shed roof, was used for making double and single laths.

All of the mills for grinding products had millstones. These stones were very important to the miller. On a windless day he would take advantage of the calm to dress his stones. These stones marked the financial success of the miller and each had his own tricks for cutting the stone to get the most out of the grain. No two stones were ever alike and the number and angle of the curves in the giant stones and their sharpness were tightly guarded secrets. With the shutters drawn and a softly glowing oil lamp to cast the proper shadows on the design cuts, the miller would toil away with hammer and chisel—and always a watchful eye for a neighboring miller hopeful for a glimpse of his more successful colleague's dressing techniques.

Around 1850 approximately ten thousand windmills dotted the polders and crowned the walls of the Dutch villages, the largest number that ever existed. After that time the number of mills decreased steadily, at first slowly, later on more quickly.

After Newcomen discovered the concept of the atmospheric steam engine in England in the eighteenth century, steam power was used to drive the pumps in English coal pits. Then came the condensing low-pressure steam engine invented by James Watt. These things led to the ultimate triumph of steam engineering. At first ships were propelled by steam power and later, by pumping stations. This concept was first applied in Holland in 1825 for the reclamation of the *Zuidplaspolder*. The idea was slow in taking hold at first, as coal consumption was high and expert enginemen were scarce.

The decision to use three large steam engines to reclaim the *Haarlemmermeer* (Haarlem Lake) turned the tide. From 1848–1852 these engines drained the large lake, and Holland too was convinced that steam must replace the wind. Water levels in the polders could be controlled faster and more efficiently by steam than by windmills. Thus the landscape gradually changed as the boiler house, its pointed gables and arched windows with the tall brick

72

Modern engineering has replaced the drainage windmill.

chimney and curling black smoke, gradually replaced the familiar graceful wind-filled sails. By 1900 only four thousand of the faithful giants, deprived of their usefulness, stood motionless as the wind blew aimlessly across the polderland.

Natural calamities were a partial cause for the disappearance of windmills. Fire from lightning or braking on a runaway mill are examples. Storms, too, took their toll, but many more were torn down and others disfigured to be

73

made over as part of steam pumping complexes.

The latter half of the nineteenth century saw the same fate for the industrial windmills as for the polder mills. Steam was faster. Then came the internal combustion engine and the electric motor in the early twentieth century and many more windmills fell victim to the massacre. Today only 957 remain. These windmills, which saw so many generations come and go, stand as idle memorials incarnating the spirit of the toil, joy, and grief of those past generations.

Chapter 4

Windmills in America

In early versions of the official seal of the state of New York, it was a windmill around which the seal design was made. The selection of a windmill for this purpose is appropriate, for it was on Manhattan Island in 1633 that the Dutch were believed to have erected a wind-powered saw-mill, the first of literally millions of windmills that would play an important role in the development of America.

The windmill idea spread to Long Island where these tireless machines soon brought life to the barren shoreline. Most Long Island mills pumped seawater for manufacturing salt, but that was only the beginning of a wide variety of tasks assumed as windmills began moving westward in a new nation born in revolution.

Other windmills made flour, tobacco, mustard, or sawed wood. Windmills were America's first weather vanes. Sailors of the day watched the direction of sails set on coastline windmills, then adjusted their own sails to match. On Long Island, ferryboat operators often based their sailing schedules upon weather information related by windmills. "Operating daily except when the windmills have lowered

An old print of New Amsterdam in the seventeenth century

their sails" was a common poster read by ferry passengers.

Though picturesque, the first windmills brought to America by colonists were cumbersome and costly. Often, they did not fit the needs of the settlers of the New World.

In the Old World, man had plotted for centuries how best to harness the ofttimes violent, uncertain, natural force of wind. In Europe, he succeeded by designing and building large windmills, but in America, the winds were different.

So the early colonists were challenged to take this European giant and trim it to the delicacy and perfection of action necessary to adapt to the conditions and needs of the New World.

Through the years, as this challenge was met, the appearance and operation of the windmill in America differed considerably from its Dutch ancestor. The main difference

76

was in the form of the wheel which received the impulse of the wind. Instead of the small number of large cloth sails of European mills, the American wheel consisted of a great number of short, narrow blades. At first these were made of wood, later, of steel. With this change sufficient wind surface was provided, and size, capacity, and strength were obtained with a minimum of weight as well as symmetry and convenience of construction.

American windmills, mostly of the vertical variety, progressed into several types. American windmills differ among themselves, principally in the way the wind-driven wheel is constructed and how it operates, and in the type of governor, or device, that controls the speed of the wheel. (Like many early horizontal mills, the surface which catches the wind is called a wheel instead of a sail.)

One of the two principal types of American windmills utilizes a wheel that has slats or blades which are flexible or folding; they can be regulated open or shut like a venetian blind, depending upon the strength of the wind. To control the slats, this type employs pivoting weights called a centrifugal governor. As the wind increases, the wheel turns faster, and the weights fly outward and upward from their pivot. The pull of gravity becomes stronger the higher the weights fly, and this tends to pull the weights down, altering the position of the slats. By using just the right amount of weights in the governor, the slats open and close to control the speed of the wheel.

In the second major type, the slats are fixed and cannot be opened or shut. To control the speed of this windmill's wheel, a vane is attached directly behind the wheel; as wind pressure increases, the vane turns the wheel away from the wind and thus slows the wheel's speed.

Both types of windmills are also affixed with a rudder positioned behind the wheel. The plane of the rudder is perpendicular to that of the wheel, so that the winds act directly upon it. The rudder thus controls the direction the wheel faces, turning the wheel as the wind shifts.

Besides these two principal types there were other less popular ones. In a third type, a solid wind wheel was used, but the regulation was effected by placing the rudder at a slight angle to the center line of the shaft so that the mill was never entirely normal to the direction of the wind.

In a fourth type no rudder at all was used. The pressure of the wind on the wheel alone was relied upon to bring the wheel into the proper direction.

The purposes for which wind power was used were many, and the manufacture of windmills became one of the important industries of the United States. Many thousands of windmills were produced annually for a trade as broad as the country and later reaching far beyond for a great foreign demand.

As the early settlers moved steadily west, the windmill moved with them. It was a machine that could do much and asked little. Windmills pumped the water that enabled the railroads to span across the continent and the cattlemen and farmers to thrive where otherwise life would have been unbearable or even impossible.

The builders of the first mills were millers themselves. However, as wind power was first the problem of the sailor, boat builders became the experts at windmill design and

A typical modern American windmill in Southern California, made by the Eclipse Wind Engine Company. Rudder-like vane at left (top of windmill) controls direction windmill is pointed.

sail makers made the canvas vane covers. During the nineteenth century carpenters and machinists began specializing as "millwrights."

The Halladay "standard" windmill is generally regarded as the pioneer of the open or sectional windmill. It was invented by Daniel Halladay and John Burnham.

Burnham was in the pump business, and his product was a hydraulic ram. What he needed was a motive power for the equipment he sold. The idea of using windmills came to him as he listened to the ceaseless winds coursing over the Western prairie. To him, the wind seemed as if millions of horses were sweeping through the heavens, visiting every farm in the known world, offering an energy that could be utilized for the saving of human labor.

The "horses" obsessed Burnham and provided the inspiration that led to his windmill design. There was no factory anywhere in the world at this time making the kind of self-regulating windmill Burnham wanted. He knew why there was such a difficulty in producing a machine that could withstand the strong prairie winds, and he felt that if this difficulty could be overcome he would find certain success.

Recognizing his limitations as an inventor, he sought help from Daniel Halladay, who ran a machine shop in his village. After much discussion Halladay said, "I can invent a self-regulating windmill that will be safe from all danger of destruction in violent wind storms. But after I make it, I don't know a single man in the world who would want one." Burnham assured him that this was not so, that experience had taught him there was a great demand for such a windmill. The two formed a business partnership and in 1854 organized a joint stock company in South Coventry,

Connecticut, which grew into a manufacturing giant, producing millions of windmills and earning millions of dollars.

The invention of the solid-wheel windmill in America is equally interesting. Its inventor was a missionary named Wheeler who was patiently laboring to bring Christianity to the Indians in Wisconsin. In 1844, an idea occurred to him. Why not, he asked himself, use a windmill to grind corn and pump water, two tasks the Indians were laboriously doing by hand.

For twenty-two years, the idea lingered in Wheeler's mind, but he found himself too busy to translate it into action. In 1866, however, when a wrist injury resulting from a fall from a ladder forced him to slow down, the windmill proved a means of keeping both his hands busy and his mind occupied.

Ironically, Wheeler's son had fallen out of a tree on the same day, breaking a leg. Father and son decided to turn their temporary handicaps in their favor; after all, the elder Wheeler still had sound legs, the younger, stout hands.

Patiently, they worked together on a new windmill, chiseling the wood parts of their mill, then hiring a village blacksmith to make the metal mountings. On April 26, 1866, the new invention was put into operation. At first it worked well, but was later blown to pieces by a hard storm. Wheeler then conceived the idea of the side vane set against the wind, so that a very strong wind would blow it around at an angle out of the wind. In two months the self-regulating mill was in operation. In the spring of 1867 a patent was granted and manufacturing was begun. This was the foundation for the Eclipse Wind Engine Company which manufactured windmills by the millions for years to come.

The most revolutionary invention in the windmill industry was the introduction of the steel back-geared mill, named the Aermotor. A lesser number of wide steel slats, curved in cross section, replaced the thick wooden slats common to the old-style sectional open wheel mill. A steel windmill was able to run in very light winds with efficiency where the old heavy wooden wheel could not respond. These steel mills were galvanized to insure durability and many can still be seen today quietly performing their tasks.

Many a farmer of this era, not always able to come up with the cash needed to buy a windmill, made his own. Most of these inventions of necessity were "jumbos"—mills fashioned like an overshot water wheel with the wind taking the place of the water in turning the blades. Even an axle from an old wagon, bolted, hub and wheel, to the side of a barn, with fans nailed to the spokes, served admirably.

The popularity of the windmill brought a new profession to the West. Generations of windmill men and well drillers came across the plains from the Dakotas to the Pecos.

Windmill builder Jake Friesen erected four thousand windmills in a quarter of a century across southwest Kansas and the Oklahoma Panhandle, training his twelve children to follow in his footsteps as he went.

A cowhand learned quickly that with a little mechanical skill he could raise the $30 a month he earned from wrangling into $126 a month as a windmill boss.

In its heyday, perhaps a hundred years in this country, the windmill performed many services and paved the way for other conveniences.

For the farmer and his family it put running water in the farmhouse for drinking, washing, bathing, and chilling the

Mill grindstones are often used to decorate places.

cream and butter. On the land it watered the cattle, greened the pastureland, and grew the fruit trees and flowers.

It served as a weather indicator and a lookout tower from which strayed cattle could be spotted. In the harsh West, the windmill occasionally proved the means by which a destitute farmer, no longer able to face the privations of the frontier, could end his despair: towering high, it made an ideal makeshift gallows.

Wherever you go across America you will find millstones. They serve side by side for a garden walk, stand as monuments, or decorate a country inn. While most of the old wooden mills have long since fallen to rack and ruin, their sturdy grinding stones still remain.

These millstones, dressed with cut grooves, turned one upon the other to crush the grain and spilled it off the edges of the stones. Because of the resemblance to plowed farmland the grooves were called "furrows," and the smooth surface of the stone was called "the land." Many a barn hex sign or early settler's patchwork quilt was inspired by his millstone design.

There were few more important figures in American business than the miller. The miller was always the link between farmer and industry, whether the reason was timber to be cut, salt to be made, or flour to be ground. All of these were tasks windmills could perform. The miller became a price setter, counselor, buyer, seller, and often banker. His advice on business and banking matters was sought and paid for, as the services of a lawyer might be. His earnings were primarily tolls collected for milling. Cash was seldom used in those days, so a portion of the grain he milled might serve as payment.

The first toll at Plymouth was set at four quarts for each bushel of corn ground. In 1824 an act in the Statutes of Connecticut gave the miller three quarts of grain for milling each bushel, one quart for each bushel of malt, and one pint for each bushel of meal. If the miller took more than his share, he was fined two dollars—one dollar for the owner and one dollar to the town in which the incident took place.

Like the miller's European counterpart, a careless miller's life was always a short one. He could be knocked in the head by a spar, thrown from a whirling sail, or caught and ground up in the gears. "Killed at His Mill" was often the

An American windmill made by Dempster Company, Beatrice, Nebraska, but operating in Niger, Africa.

miller's epitaph. The millstones in use when a fatal accident occurred were considered unlucky and usually were used to mark the miller's last resting place.

The many types of mills and their amazing number were a part of the American picture that is often forgotten today.

At the turn of the century, the production of windmills

in America was a $10 million a year industry. In 1935, a score of United States factories still made 100,000 mills a year. In 1962 figures of the Department of Commerce showed only 6,484 windmills manufactured in the United States. Today the domestic windmill market is divided between two firms, Aermotor, Inc., of Chicago, and the Dempster Mill Mfg. Co. of Beatrice, Nebraska.

Over the years, the windmill market has continued to decline because of rural electrification. Most of the sales are in three relatively small, distinct areas: the cattle-grazing area of southwest Texas, the panhandle area of Texas and Oklahoma, and the cattle-grazing area of western Nebraska. American windmills once enjoyed booming foreign sales in some of the underdeveloped nations. Later, South Africa, Australia, and Argentina began making their own mills, confining American sales to a small export market in Mexico, Central America, and South America.

Reminders of the windmill's growing obsolescence can be seen from roadsides across America. Some, silhouetted against the sky, are stilled for all eternity. Others, their heads drooping, are like the wilted blossoms of the sunflower on a summer prairie. Saddest of all, perhaps, is the tower, shorn of its graceful blades, that is laced with electrical wires and crowned with a television antenna.

Chapter 5

Windmills of Tomorrow

When Mr. and Mrs. Henry Clews built a home in the Maine woods not many years ago, they were astonished to learn how much money the local power company wanted to install the equipment necessary to provide electricity. Clews decided the cost was more than they could afford, but Mrs. Clews made it clear that she did not want to power the new home with kerosine lamps and candles.

Mr. Clews is a former high school teacher whose hobby is puttering with mechanical things. He decided to "electrify" the home himself—using a windmill. He bought a small windmill in kit form from an Australian company, along with a forty-foot high steel tower. Working in his spare time with friends to help him, he soon had the windmill (a wind *generator*, actually) in place and operating.

Today, Clews' homemade wind generator is capable of lighting eight 75-watt bulbs at one time, and of running a television set, stereo, electric typewriter, blender, toaster, vacuum cleaner, portable saw, electric drill, and even a water pump.

The installation of the operation cost far less than the

John Bennett, an engineer with Sun Oil Company, in front of a windmill he built in Texas to provide power for his home.

electric company would have charged the Clews. And since the source of energy—the wind—is free, the only monthly expense involved is for maintenance, which is very low.

In contrast to the way the Clews built their windmill to power a single home, there are revolutionary planners like Professor William E. Heronemus of the University of Massachusetts who envision vast windmill systems that could provide power for wide regions of the country.

Dr. Heronemus, a professor of civil engineering, foresees as one example a network of windmills floating off the coast of New England, each about three hundred feet high. These wind generators would electrolyze seawater into oxygen and hydrogen. The oxygen would be separated from the hydrogen which would be stored in huge tanks on the sea floor. When it was needed, the hydrogen would be shipped ashore to be used in energy fuel cells. Professor Heronemus maintains that such a proposed windmill system could take over the entire task of providing electricity for six New England states.

Between the two extremes, there is a flurry of research and development these days on the comeback of the windmill. The energy crisis of a few years ago and rising costs of energy have suddenly made man's oldest machines look good again. But the machines involved are not ones for grinding corn or pumping water, but for generating electricity.

Electric power generation is a relatively new role for the windmill. Less than a century ago, in 1890, a man named LaCour built a mill in Denmark that used wind power to generate electricity. But it was found that watermills could do the job better, and the coal-burning or wood-burning turbine better still. For a long time after LaCour's invention, windmills as energy deliverers were used only in places where there were no streams to be dammed or where there were low supplies of fuel.

But there were isolated exceptions. In 1894, only four years after LaCour's mill, explorer Fridtjof Nansen designed such a unit out of necessity. At a time when most major cities were still relying on gas and kerosine for heating and light, his wind-driven dynamo charged batteries for electric lights in the polar wilderness.

The operating concept of such a unit is quite simple. The wind turns a propeller or vane, which is attached to a shaft. The shaft rotates and, either directly or via a series of gears and couplings, spins the rotor of a power generator, which in turn feeds electrical current into a transmission line or storage unit for eventual consumption.

Wind does not blow with the same strength all the time, of course, and sometimes it does not blow at all. A windmill cannot produce electricity in winds below six miles per hour. So the problem of how to store up the electricity generated on windy days for use on windless days is one of the most critical in wind-energy research. A second problem is cost; until better and more economical wind generators are developed, those designed to provide substantial amounts of energy are quite expensive to build, although very little cost is involved in their actual operation.

Though elusive, wind power is nevertheless enticing, because it is continuously regenerated in the atmosphere under the radiant energy from the sun. It is therefore a self-renewing source of power, as opposed to energy from fossil fuels.

A wind generator, wind turbine, or aero generator (the terms are interchangeable) has already been used to deliver commercial power. From 1941 to 1945, the years of World War II when electricity was in great demand, the largest windmill in history was built on a mountain called Grandpa's Knob near Rutland, Vermont. The site was chosen because it rose two thousand feet high; as we have seen, the strength of wind increases with height off the ground.

Built by the Central Vermont Public Service Corporation, the turbine sat atop a 110-foot steel tower. Its stainless steel blades were more than eleven feet wide and nearly seventy feet long. The blades (or "sails" in windmill language) drove a generator that converted the wind's energy into electricity.

So strong was the wind generator that it once withstood a storm with 115-mile-per-hour winds without damage. In one test, it ran for more than 143 continuous hours. Off and on, its 1,250-kilowatt generator pumped into the surrounding area enough electricity to power as many as two hundred homes. But even in the 1940s its construction and maintenance cost was considered extremely high; when one of its giant arms broke off in 1945, the wind generator was shut down for good.

American wind-power experiments came to a virtual halt after World War II. After all, "cheap" fossil fuels were available once again, and only a few people could foresee the day when their cost would spiral as their supply began to run short.

In June, 1973, a Wind Energy Systems Workshop was held in Washington, D.C. It was the first step in a major five-year program set up under the guidance of the National Science Foundation (NSF) and the National Aeronautics and Space Administration (NASA).

A 100-kilowatt windmill will be built from this model at Sandusky, Ohio, in a wind-energy experiment sponsored by the National Aeronautics and Space Administration and National Science Foundation.

NASA-National Science Foundation experiment wind-generator now in operation at Plum Brook, Ohio.

As a result of the federal programs, the investment in American wind-energy research programs soared from $200,000 in 1972 to more than $12 million four years later. Many universities as well as private corporations and individuals have now joined in the search for better ways to generate electricity by using the wind.

NASA estimates that if the interest continues, as much as 5 to 10 percent of the country's energy needs could be filled by wind generators by the year 2000. Toward that goal, NASA recently began operating its own experimental wind turbine generator at Plum Brook, a test site near Sandusky, Ohio. It is a 100-kilowatt windmill that reaches a height of more than 100 feet. It has two blades, joined at a hub, that span 125 feet. When fully operational, it is expected that the Ohio generator will fulfill all the electricity requirements of at least thirty homes. It is designed primarily as a prototype, however, a smaller version of what wind generators eventually could become.

On a large scale, wind power theoretically could produce huge amounts of electricity. Another proposal by Professor Heronemus, for instance, envisions some 300,000 wind turbines on the Great Plains. They would be spaced as closely as one for each square mile. Each 850-foot tower would carry twenty turbines, consisting of a two-bladed, fifty-foot diameter propeller. The Great Plains network, according to Professor Heronemus, could provide the equivalent of 189,000 megawatts of nuclear power plant installed capacity.

Professor Heronemus adds that an even greater capacity is possible, but that it could not be met over a year's time because of the unpredictable nature of the wind. He worked out the comparison with nuclear plants mathematically to provide a more realistic measure of the electrical production that could be anticipated from wind power. Considering that the *total* capacity of all electrical generating plants in the country in 1971 was 360,000 megawatts, the Heronemus Great Plains plan would become a significant supplier of power.

As mentioned earlier, winds are produced by solar energy. The differential heating by the sun of the earth's surface causes a lateral heat flow that keeps the particles of the global atmosphere in motion. Still other winds are caused by the *evapo-transpiration cycle*—heat being stored in the atmosphere until it rains and cools things off, allowing the heat buildup to start all over again.

There is a theoretical limit to how much wind is available worldwide at any given time, which wind generators could tap and transform into electricity. But it is such a vast supply that man probably never would be able even to approach tapping all of it.

The World Meteorological Organization (WMO) many years ago attempted to estimate how much kinetic energy in the atmosphere is potentially available to windmills on earth. The organization calculated that 20 *billion* kilowatts of power would be available for "harvest" at a series of carefully selected windmill sites around the world. These would be mostly on treeless mountain peaks and shoreline cliffs in areas known to have a very high average wind velocity.

Other estimates of available wind energy range as high as 80 *trillion* kilowatts in the northern hemisphere alone. How does that translate in terms of power the world now uses? According to the Federal Power Commission, in 1970 the total world electrical generating capacity was one billion kilowatts, of which the United States alone generated and used nearly one-third. In other words, if the WMO's estimate is correct, windmills at only a few special locations theoretically could produce twenty times the world's then-current demand for electricity.

Not even the most optimistic of "wind engineers" suggest that windmills could be an answer to all the world's energy demand. But there seems little doubt that as a supplier of *supplemental* power, windmills have a bright future.

Unfortunately, this potential energy supplement is not available equally around the world, because winds are not equally distributed. The wind is quite consistent, however, within large areas, and for this reason certain winds have been given names according to their location. There are trade winds, for example, as well as doldrums, westerlies, and easterlies.

Generally speaking, the major wind systems of earth intensify from equator to either pole. Their intensity also varies with the relationship of land and water masses, topographic features, and the like.

It is well established that the wind blows especially hard at high elevations; gnarled, stunted trees on high mountaintops are evidence of this.

Whatever its strength, however, there is a theoretical maximum limit to the amount of energy that can be extracted from any airstream. Scientists estimate this limit at 59.3 percent. But there is another scientific formula which helps the windmill builder determine the type of construction and length of the blades or sails which would best take advantage of any particular wind.

The formula is that the wind power varies as the cube of

velocity and as the square of any circular area through which the wind passes. Let's study one example in simpler terms to see what this means. A twenty-foot diameter windmill blade in a fifteen-mile-per-hour wind will yield 2.55 kilowatts of electricity. But a forty-foot blade—only twice as long—will yield 10.23 kilowatts in the same wind, or four times as much.

Many of these winds are generated principally at sea. The ocean, therefore, was regarded as a prime source of wind-derived electrical power when Professor Heronemus suggested his ocean-going windmill plan a few years ago.

As he conceived the idea, the windmills would be mounted on floating platforms or concrete-pile towers like those oil companies use to explore for and produce offshore oil. They would be located on the Georges Banks and on Nantucket and New York shoals. The windmills would be used to generate electricity to run the offshore electrolyzer stations mentioned earlier.

The network would include eighty-three wind units, compressor and deep-sea storage systems, an offshore collection subsystem, and shoreside terminals, distribution subsystems, and fuel cell substations. Each wind unit would comprise 164 wind stations (each with three two-bladed windmills) arranged in concentric rings around an electrolyzer station.

The cost of the entire system would be about $22.3 billion, according to Professor Heronemus. That's many times higher than the cost of developing conventional electricity-generating systems for an equal output of energy. However, as Dr. Heronemus contends, it is about equal to costs of developing nuclear energy, another developing source of energy as fossil fuels continue to dwindle.

It is not only New England and the Great Plains that have attracted Dr. Heronemus' attention. He believes that the entire northern third of the country has winds strong enough and consistent enough to produce electricity.

Meanwhile, on the opposite side of the country, researchers at Oregon State University have been studying the power potential of winds on that state's shoreline. "If it were possible to install additional generators in dams to use water pumped back into reservoirs," says Dr. E. Wendell Hewson, "you could use the existing system without building more dams and further endangering the environment. Raising the level of a reservoir by just a few feet could make quite a difference for hydro-generation."

Wind-energy research is not limited to the United States. Denmark, for instance, had developed a system of small windmills with a total generating capacity of 100,000 kilowatts prior to World War II. During the German occupation, these mills became a main means of lighting and heating homes. Many of the old Danish electrical mills are still working today.

Wartime Germany conducted windmill research, too, after discovering that even though Paris is not a windy city, the winds atop its Eiffel Tower were constantly quite strong. This discovery gave the Germans, who had occupied Paris, an idea of how the most efficient wind generators should be built. By 1944, Germany was preparing a series of extremely high towers with 400-foot diameter propellers capable of generating ten to fifty megawatts of electricity. As Germany began to lose World War II, however, steel became in such short supply that the project was abandoned.

By 1950, a number of English scientists and engineers

sought ways to combine wind and tidal power as a means of generating electricity. In postwar England, fuel was scarce and the cost of developing the windmills seemed feasible. But research proved that even though the winds off the coast of Wales and Cornwall seemed ideally suited for harnessing, there were too many problems to overcome, and the project was cancelled. Other nations that have engineered windmill programs include South Africa, Argentina, Israel, India, Hungary, and the United Arab Republic. The success of these ventures has varied, but in recent years, the pace of research and development has quickened worldwide.

It is not only national governmental agencies that have developed wind-energy programs. Universities, private companies, at least one city, and many individuals have joined the pursuit of wind energy as well.

In the United States, a Southwestern college recently began a course in windmill repair; another teaches students how water-pumping windmills might be converted to those producing electricity.

Private corporations are developing their own wind-generating systems for specific tasks. One company now manufactures a small wind generator designed specifically to supplement electrical power on offshore oil platforms. And a firm in San Antonio, Texas, has proposed a way of using an abandoned mine to store electricity that windmills first generate. On windy days, a wind generator would pump air into the empty mineshaft. On calm days when the windmill is not working, compressed air would be released through gas turbines to produce needed power.

And in 1975, Honolulu became the first American city to invest tax dollars in wind-energy research. So far, studies

Small wind-driven generator used to power marine warning lights and fog horns near oil platforms in the Gulf of Mexico. Simple system replaces battery-operated warning system which cost $1,500 per year.

Called a Vertical Axis Windmill, this device is under study at NASA's Langley Research Center, Hampton, Virginia. Wing-shaped airfoil rotates in any direction to convert wind power to electricity.

there have concluded that a workable wind generator could be built for $50,000 that would last perhaps fifty years. It would produce about as much energy each year as $7,000 worth of oil or coal—thus paying for itself in seven years.

Lastly, there are the world's individuals—the amateur tinkerers and inventors—who have joined the governments, private corporations, and universities in harnessing the capricious wind. Today, small, inexpensive windmills are powering everything from automobiles to homes; perhaps their greatest use is in remote areas like that in which Mr. and Mrs. Clews live, where conventional power supplies are costly.

The Clews bought their windmill in kit form, but many others have been assembled from junk parts. For a few hundred dollars and a little mechanical knowledge, windmills are supplying many homes with electricity to run stereos, lights, heaters, and television sets. And there is a never-ending supply of "fuel"—the wind.

Bill Gibbons of Ontario, Canada, built a six-volt windmill from parts of an old car and other junk parts, for only $100. Near Albuquerque, New Mexico, several families enjoy television, stereo, and other electrically powered appliances for which the source is a combination of windmills and solar energy. Developed by an inventor named Robert Reines, the first wind unit was installed in 1972 and two others were added later. Today, they generate a total five kilowatt capacity; while the sun provides the source of heating the homes, the wind—harnessed via the windmills —powers everything else. The system has proved so efficient that Reines found a way to pump leftover electricity into a storage unit which he uses to run a small, four-wheeled car. And in Farisita, Colorado, Winnie Red Rocker built a workable electrical windmill from nothing more than inexpensive hardware and parts donated by friends.

As interest in tomorrow's windmills soars in an energy-hungry world, what of the windmills of yesterday? What of the colorful old post, smock, and tower mills of Europe that performed so faithfully for generation after generation of millers and to which the modern windmills owe their origin?

Sadly, perhaps, in most countries there are almost none left standing, and what they once looked like and how they operated remains answered mainly in history books.

In two countries, however, the decline of the historic windmill was halted almost at the hour of their final destruction. In Holland it became apparent by the early 1920s that the steam engine was on the verge of eliminating the country's national symbol altogether. Electricity had become cheap; the old windmills just couldn't compete economically anymore in the tasks of pumping water, draining fields, and grinding grain.

Alarmed at what they saw happening, a group of private citizens in 1923 formed an organization they called *De Hollandsche Molen* (literally translated, "The Dutch Windmill") as an attempt to meet the increasing competition on equal terms. The society set two goals. One was to select a number of windmills which would be "upgraded" by installing modern equipment, yet which would still function as they always had in their varied tasks. In other words, metal parts might be substituted for wooden ones, or the buildings themselves strengthened with modern materials.

Related to this was a search for other windmills that could be restored, in working condition, exactly as they once had been, but with original parts. Where original parts could not be found, new ones would be made, but of exactly the same material as the originals. These mills would operate as they once had, not necessarily to grind grain or pump water on a commercial scale but to demonstrate to visitors how they once performed these jobs.

The job undertaken by *De Hollandsche Molen* proved tremendously expensive, so in 1924, at the request of the society, the Dutch government informally accepted the existence of windmills as historical landmarks and agreed to contribute small sums toward their maintenance. In that year, too, every local town council was sent a copy of a circular from the national Minister of Education, Arts, and Sciences stating that no windmill be demolished without first consulting the society to see if it had restoration possibilities.

Through the years, the society—consisting mostly of volunteers—restored one windmill after another. The work became even more significant during World War II when some of the old mills fell victim to military attacks.

The government gave Holland's windmills an even more formal blessing in 1945 with passage of the National Windmill Act. It strictly forbade the destruction of *any* building in the country that had a potential historic value. A great many of the buildings so named were windmills, and today there are more than nine hundred fully restored, working windmills that stand as a result of that act.

In 1967, Holland went one step further. A Guild of Volunteer Millers was formed on a national scale. Students interested in their country's windmill history are sent to a special school where they learn the ancient trade of the miller. At the end of the course, all are tested by experts. Those who pass are given certificates allowing them to op-

Fantail and rear side of Saxtead Mill, England, one of England's best-restored mills, and open to visitors year-round.

erate windmills. As volunteers, they receive no salary; their motivation is entirely a desire to keep windmills running as the way of best demonstrating their history.

In England, there are even fewer old mills still in existence. Altogether there are about one hundred. They are found in many parts of the country, but the greatest concentration, and some of the best examples, are in southeast England and the East Anglia region. Many are maintained by public bodies or private associations. Some are privately owned, and many are open to the public on designated days throughout the year.

The windmill restoration effort in England began with

Smock mill in Denmark. Sails were damaged in a storm.

a concern of individuals, and later was given formal status by the Society for the Protection of Ancient Buildings, a government agency.

While many of the English windmills are still in working order, others have been converted to other purposes. Perhaps the most unusual is one twenty-two miles south of London at Reigate, in the county of Surrey. Built three centuries ago in the reign of King Charles II, it is now used as a church, and services are held there once a month.

Of most of England's windmills, as with Holland's, it can be said that while their working days are over, they are still useful, delightful complements to the countryside, fine studies for the artist and photographer, reminders of a bygone age when wind was a major source of power.

Today, that "major source of power" is enjoying new attention, and though the modern electricity-generating windmill bears only faint resemblance to its historic predecessors, it enjoys a close kinship nevertheless.

Perhaps it was former Secretary of the Interior Stewart Udall who once best summed up the unchanging worth of windmills both old and new:

"Windmills are much more than relics. They are symbols of a sanity in a world that is increasingly hooked on machines with an inordinate hunger for fuel and a prodigious capacity to pollute. Ecologically, the windmill is one of the few perfect devices. It harnesses a complete free resource to pump water or generate electricity under conditions that respect the laws of nature."

Index

107

109

The Authors

Residents of La Jolla, California, Joseph E. and Anne Ensign Brown frequently collaborate on nature and science subjects. Joseph Brown has authored seven books, including three previous Dodd, Mead children's books. His wife formerly wrote for and edited oceanographic journals and has illustrated children's books. The Browns share a sailor's appreciation for the wind aboard their 28-foot sloop, *Whisper*.